Assessment of challenges and opportunities of traditional soil conservation practices in Dabre work woreda, East Gojam Zone, state Ethiopia

Amanuel Berhanu

Bibliographic information published by the German National Library:

The German National Library lists this publication in the National Bibliography; detailed bibliographic data are available on the Internet at http://dnb.dnb.de.

ISBN: 9783346718969
This book is also available as an ebook.

Nymphenburger Straße 86
80636 München

Print and binding: Books on Demand GmbH, Norderstedt, Germany
Printed on acid-free paper from responsible sources.

GRIN web shop: https://www.grin.com/document/1020380

Assessment of challenges and opportunities of traditional soil conservation practices in Dabre work woreda, East Gojam Zone, state Ethiopia

Author: Amanuel Berhanu (MSc)

Department of Rural Development and Agricultural Extension; College of Agriculture and veterinary Medicine, Jimma university; Ethiopia

Abstract

The study was conducted in Diyateba kebele and Dijumeti kebeles at Debirework Woreda, East Gojjam zone. The major purpose of the study was to assess the opportunity and challenges of traditional soil conservation practices in Diyateba kebele and Dijumeti kebeles kebele. It also among at investigate the opportunities and challenges of traditional soil conservation practices. The data for this study were collected from both primary and secondary sources by using simple random sampling techniques. The primary data was collected through questionnaires and interviewing farmers who have long history and more experience about the data. The secondary data sources were taken from books and Diyateba kebele and Dijumeti kebele Administrative office. The study should that the farmers of the study area have a good awareness of the negative effects of soil erosion on socio economic and environmental aspects of the society. To overcome the negative effects, farmers of the study area have practice several traditional soil conservation measures. These practices are both biological (application of manure, crop rotation) and structural measures (soil or stone bund) to protect soil from and maintain soil fertility. To apply those traditional soil conservation practices there are opportunities like distance from homestead, presence of livestock ,area of cultivated land and absence of inorganic fertilizer. And has challenges like socio economic, farm institutional and technological challenges. To reduce these problems there should be improved knowledge of local communities 'towards traditional soil conservation practices.

Key words: *Traditional conservation practices, stone bund, animal dung, crop rotation.*

Table of Contents

Abstract i

1. INTRODUCTION 3

1.1 Background of the study 3

1.2 Statement of the problem 4

1.3. Objectives of the study 4

1.3.1 General objective 4

1.3.2. Specific objectives 4

1.4 Research questions 4

1.5. Significance of the study 4

1.6 Scope of the study 5

2. LITRATURE REVIEW 5

2.1. Concepts of soil conservation and soil erosion 5

2.1. 1. Soil Erosion 5

2.1.2. Soil conservation 5

2.2. Soil conservation in Ethiopia 6

2.3 Traditional soil conservation methods 6

2.3.1: Soil bund 6

2.3.2: Animal dung 6

2.3.3 Crop Rotation 7

2.3.4. Application of manure 7

2.3.5. Crop rotation 7

2.3.6. Soil or Stone bund 7

2.4. Opportunities of traditional soil conservation practice 8

2.5. Challenges of traditional soil conservation practice 9

2.5.1 Socio cultural challenges 10

3. RESEARCH METHODOLOGY 11

3.1 Description of the Study Area 11

3.2 Research design 12

3.3 Sources and types of Data 12

3.5 Sampling Techniques and Sample Size ... 12
3.6 Data analysis ... 14
3.7 Hypothesis of Variables ... 14
3.7.1 Independent variables ... 14
3.6.2 Dependent variable ... 15
4. RESULTS AND DISCUSSION ... 16
4.1 Descriptive Statistics ... 16
4.2 Socio-demographic characteristics ... 16
4.2.1 Sex, Age and marital status of Respondents ... 16
4.3. Nature, Type and problem of erosion in the study area ... 19
4.4 Farmers perception on Cause of soil erosion in the study area ... 20
4.5 Effects of soil erosion in the study area ... 20
4.6 Traditional soil conservation methods being practiced in the study area ... 21
4.6.1: Soil bund ... 21
4.6.2: Animal dung ... 22
4.6.3 Crop Rotation ... 22
4.7 Practices of soil and water conservation ... 22
4.7.1 Traditional Cut-off Drains ... 23
4.7.2 Traditional waterways ... 23
4.7.3 Drip irrigation ... 23
4.8 Challenges of Traditional Soil Conservation Practices in the Study Area ... 24
4.8.1 Socio cultural challenges ... 24
4.8.2 Deforestation ... 25
4.9 Opportunities of Traditional soil conservation practices in the study area ... 25
5 CONCULISION AND RECOMMENDATION ... 27
5.1 CONCULISION ... 27
5.2 Recommendation ... 28
6. REFERANCES ... 29

1. INTRODUCTION

1.1 Background of the study

Agriculture is the major source of livelihood in Ethiopia; however, land degradation in the form of soil erosion has hampered agricultural productivity and economic growth of the nation (Balana,2010). Land degradation, low agricultural productivity, and poverty are crucial and closely related problems in Ethiopia (Yitbarek et al, 2012). Land degradation is a great treat to the future productivity and it requires a great effort to protect resources (Girma, 2001). To overcome the problem of land degradation on agricultural productivity, Ethiopia has made efforts to launch afforestation and conservation programs with the support of both governmental and non-governmental organizations, however, success to date has been limited (Bishaw,2001).

A study conducted by Yitayal (2004) on determinant of use of soil conservation measures by smallholders in Jimma Zone, Debo district showed that the slope of the farm plots and distance of the farm plot from residence significantly influence the use of both traditional and improved soil conservation measures. As Yitayal identified determinant of use of conservation measures the researcher becomes also identified different traditional soil conservation methods and the challenges including (social, cultural, technical, economic and farm size challenges) and opportunities in diyatebaKeeble.

A study conducted by Senait (2005) on determinant of choice of land management practice in Amhara region, North Showa zone, using multiple logistic models that leads ownership type, distance of farm plot homestead, resource availability and contact with extensions where found to be the most important factors affecting choice of land management practice such as commercial fertilizer, manure, stone/soil bunds or a combination of them. Among the identified management techniques majority are traditional soil and water conservation practices.

Most soil and water conservation planning approaches rely on empirical assessment methods by experts and hardily consider farmers knowledge of soil erosion. As a result conservation programs and approaches performed poorly (Yohannes and Herwege, 2000).

Therefore, this research proposal work is intended to contribute to find out factors that influence decisions on traditional soil conservation practice in diyatebakebele by using different scientific research methodologies.

1.2 Statement of the problem

Application of different soil and water conservation technologies is very important in addressing degradation resulting from both natural and manmade influences. There are different traditional soil and water conservation technologies in diyatebakebele through which farmers give care about their land resource. However, farmer's adoption of these technologies is much traditional and its diffusion is tradition based. Although the farmers are supported by the newly introduced calendar based soil and water conservation which is conducted around February every year, farmers are still facing problems in adopting scientifically proved soil and water conservation technologies. Therefore, there will be study which tries to find out challenges and opportunities in adopting this existing traditional soil and water conservation technologies in the keble so as to bring better conservation environment for the farm community.

Having the above base, this study is aimed to find out challenges and opportunities related to adoption of traditional soil and water conservation practices in diyateba Keble.

1.3. Objectives of the study

1.3.1 General objective

- The general objective of this study was to assess the opportunity and challenges of traditional soil conservation practices on soil erosion in Dabre work woreda, east Gojam zone,

1.3.2. Specific objectives

- To assess the opportunities of traditional soil conservation practices.
- To assess the challenges of traditional soil conservation practices.

1.4 Research questions

The research questions set to meet the objectives mentioned above are:

What are the traditional soil conservation measures which are being practiced in the study area?

What are the challenges and opportunities of traditional soil conservation method on soil erosion being practiced by the community in the study area?

1.5. Significance of the study

The result of this study is expected to contribute for the Improvement of awareness of farmers about the opportunity and challenges of traditional soil conservation practices on soil erosion in the study area. It is also expected to add for the policy makers particularly those directly related to natural resource inventories and management. The research proposal would also give general hint for the woreda office about the willingness of community towards soil conservation measure, and finally this study provides base line information for researchers, students, universities, research institutes and farmers regarding traditional soil conservation to mitigate soil erosion in the study area.

1.6 Scope of the study

Owing to time, material and financial constraints, the scope of the study is limited to the traditional soil conservation practices only in Diyateba Kebele. On the other hand, the study is not containing investigation about other practices of soil conservation practices. The study only focuses on application of manure, crop rotation and soil or stone bunds.

2. LITRATURE REVIEW

2.1. Concepts of soil conservation and soil erosion

2.1. 1. Soil Erosion

Soil erosion hazard associated with agriculture in tropical and semiarid areas and is important for its long term effects on soil productivity and sustainable agriculture .It is ,however ,a problem of wider significance occurring additionally on land devoted to forestry ,transport and recreation .Erosion also leads to environmental damage through sedimentation ,pollution and increase flooding .The cost associated with the movement and disposition not sediment in the land scape frequently those arising from long term loss of soil in eroding fields (Morgan, 2005). Accelerated soil erosion is one of the major constraints to agricultural production in many areas .Therefore, sustainable and renewal resource management practice needs to address the widespread land degradation, decline of soil fertility, unreliable rain fall and even desertification, in a context of global climate change (FAO and World Bank, 2001).

2.1.2. Soil conservation

Biological and mechanical soil conservation practices are currently applied by farmers in Iran, including, terracing, mulching, cover crop cultivation, integrated cropping and timely use of fertilizer (Refahi 2oo4 and Tavakoli and Piri, 2001). The soil and water activities were carried out using food aid in the form of food for work (FFW) .however ,free community labor was mobilized as a people's awareness increased (Badege ,2001).

2.2. Soil conservation in Ethiopia

Soil and water conservation in highlands of Ethiopia focused both on mechanical and biological measures (Babulo,et al,2009). The mechanical measures include construction of bund, terraces, diversion ditch, check dams and hillsides terraces. The biological measures comprises enclosure of degraded land from human and animal interference, tree seedling production, agroforestry tree seedling planting ,on farm lands, afforestation, and tree planting at the homesteads and in enclosures as tree enrichment (Nassen et al,2009; Mekuria et al, 2011).

2.3 Traditional soil conservation methods

Traditional soil conservation measures appears indispensably crucial and can best be understand as farming practice that evolved over the course of time through periodic experience, observation, political and ecological conditions (Rejj et al,1996).The diyatebakebele people are known by traditional soil conservation practice. As noted by the kebele elder of diyatebakebele began such practice in the earliest time in older to combat soil erosion since people have been practicing to protect the soil from loss.

There are also known as by its biological and structural soil conservation methods ,the biological methods are used to maintain soil fertility, such methods are crop rotation and application of animal dung .While the structural methods the one that mainly used for the purpose of controlling soil removal ,such as soil or stone bund, contouring.

2.3.1: Soil bund

Soil bund is an embankment constructed from soil along the contour with water collection channels or basin at its upper side. These methods are used not only for control of erosion but also for growth of grass because of the farmer's plant at the bund. soil bund in the study area applied only which is 18.2%, and 81.8 % were non use this traditional soil conservation practices this is due to the farmers complain of farm size challenges that means the farmers said it becomes narrowing of the farm land, and economic challenges (those farmers who lives in diyatebakebele were not give consideration for traditional soil conservation because they require off farm incomes rather than on farm incomes).

2.3.2: Animal dung

Animal dung is widely accepted organic fertilizer in the society of diyatebakebele. Therefore, the community of the study area often used the natural fertilizer to increase the fertility of the soil thus it can possibly be said that natural fertilizer is widely accepted and widely used for stabilization and increment of soil fertility and moisture of soil among the farmers in the kebele. According to the field survey made for this study, about 46% of farmers living in this kebele adopt application of manure. Besides presence of greater proportion of female to male population in the kebelewill contributed its own influence for the greater adoption of animal manure since female have greater relation to cows and their dung. Animal dung is mostly applied in the study area most likely due to there is sufficient amount of livestock size including cattle, sheep, goats, horse and hens, the community of the study area often used the natural fertilizer to increase the fertility of the soil this natural fertilizer is widely accepted and widely used for stabilization and increment of soil fertility and moisture of soil. Also the presence of greater average household size, in this kebele is also expected to contribute some for adoption of greater adoption of animal manure in the kebele since presence higher household size would yield greater labor size. This could possibly help for transportation of manure and other heavy labor intensive activities.

2.3.3 Crop Rotation

Crop rotation has many agronomic, economic and environmental benefits compared to mono culture cropping. Appropriate crop rotation increase organic matter in the soil, improve soil structure, reduce soil degradation, and can result in higher yields and greater farm profitability in the long term. It is used to control weeds and diseases.

2.3.4. Application of manure

Application of manure is used by many farmers in order to improve the fertility of the soil .Manure consisting of animal dung and urine and it is the best form of organic fertilizer. Farmers used manure mainly near the homestead. Farmers especially, those who were poor have increased the use of manure applied because of the high current price of inorganic fertilizers (Stark, 2000).

2.3.5. Crop rotation

Has many agronomic ,economic and environmental benefits compared to monoculture cropping .Appropriate crop rotation increase organic matter in the soil ,improve soil structure , reduce soil degradation , and can result in higher yields and greater farm profitability in the long term .It is used to control weeds and diseases.

2.3.6. Soil or Stone bund

Soil or stone bund is an embankment or ridge built across a slope along the contour .Soil bunds are made of soil or mud. On mandatory sloping areas the farmers construct the soil bunds for erosion control, on steep eroded bare lands stone terraces are mostly used structures in the study area.

2.4. Opportunities of traditional soil conservation practice

In recent years, farmers have acknowledged prioritizing invention of logistic integration of traditional realistic conservation with such as hillsides terraces, stone lines and bunds, earth contour bunds, sand terrace, trash lines, organic manuring and mulching, packaged to deliver profitable crop production stably in sufficient quantities. However, incompatible land use methods such as the modern frail conservation technologies practiced in the some soil environments will be an age long limitation to sustainable crop production (Reij et al ,1996).

Traditional soil conservation structures are not only to control run off and soil erosion as in the case of newly introducing measures, but also for the production of high quality and quantity yield for community. In Diyateba there are many traditional soil conservation practices those are both biological practice like application of animal dung, crop rotation and mechanical measures like soil or stone bund. traditional SWC practices are well recognized in Ethiopia, for example in Gojjam, ditch drain surplus runoff, in the people of Konso application Terracing have practiced on their agricultural farm land long period of time (Hurni etal.,2016:11)The farmers that live in Diyatebakebele uses these conservation practices because of there are opportunities that are suited for them . Suitable conditions that the farmers will applying these measures are;

Presence of livestock

Those farmers have large number of livestock size will more attention to traditional soil conservation practices. Because there is enough amount of dung due to presence of large livestock in each family, the opportunity of farm land for being managed properly is likely. Therefore, amount of livestock will great importance for adoption of manure as traditional soil and water conservation practices. To reduce the abovementioned environmental impacts, different production systems have been developed. Among them, organic livestock farms have been studied by several authors in order to assess their potential and impact on environmental Journal of Dairy Science.2014; and socio-economic aspects (sustainable rural development IJRSAS). 2015)

Labor

The application need extensive labor to take dung from home to farm land and it needs excessive time for collecting martial at the time of preparation. So it affects the local community who lives in study area.Agriculture is carried out, potentially changing the amount of labour required across different production stages (e.g. land preparation, weed control, harvesting and threshing). In minimizing tillage, the amount of labour required in the early stages of the growing season (e.g. land preparation) may be lower (Baker & Saxton, 2007; FAO, 2001), but may Increase labor requirements in the following stages (e.g. weed control if the same herbicides are used) (Erenstein, 2002; Pannell et al., 2014). In producing higher yields, CA may also increase the labor required in the harvesting and threshing stages (Farnworth et al., 2016)Distance of the farm from the homestead is one opportunity for farmers to apply those traditional soil conservation practices. Where the farm land is nearest to the homestead have the chance to conserve properly than that of that is located far. So in general the farm land which is far from the homestead less likely to apply those traditional soil conservation practices due to the consideration of labor.

Cost of inorganic fertilizer

The cost of inorganic fertilizer becomes increase time to time. Farmers that are present in the study area are apply traditional soil conservation practices like application of animal dung, crop rotation (of the leguminous trees that increases soil fertility) in order to reduce expense of inorganic fertilizer. The price of UREA as well as DAP is increasing from time to time. And the increment of price of inorganic fertilizer in return will decline farmer's preference of using inorganic fertilizer.The profitability of fertilizer use depends on both the technical response to fertilizers, i.e. the units of output produced from one unit of nutrient, and the relationship between output price and fertilizer price (Yanggen et al., 1998).

2.5. Challenges of traditional soil conservation practice

Although several soil conservation technologies has been developed and promoted through past decades, the adoption of many recommended measures was still minimal. So studies will need to be conducted in search of factors influencing adoption of soil conservation practice, since the 1950s a lot of attentions has been paid to the factors that determine the adoption of soil conservation practices by farmers (Ervin, 1982). Generally (Rogers, 2003) summarized the result of many adoption and diffusion studies and conclusions will provided as a means of analyzing environmental innovations.

Conservation practice adoption is a multi-dimensional process .Numerous factors determine farmers willingness to use conservation practice, some factors commonly found in the literature to be related with the adoption of soil conservation practice fall into two categories ;one individual level characteristics of the farmers including age, education, year of farming, knowledge level, awareness attitude towards conservation practice, motivation etc. Farm structure factors related to the adoption of conservation practice including farm size, income, farm profitability tenure etc.The intensity of adoption is defined as the proportion of total cultivated land that is under the CF practices as defined in this paper, hence the dependent variable is bounded the interval. We use three different models to analyses the intensity of adoption to ensure the robustness of our results: random effects Tobit, pooled fractional probit, and unobserved linear fixed effects models (Papke and Wooldridge 2008).

Traditional soil conservation practices will several benefits for the conservation of environment as a general and soil resource in particular .Even though such positive aspects and crucial value of conservation such as erosion control, maintenance of soil fertility and water conservation, but there are some limitations or challenges that will be observe in the study area This challenges may results from improper traditional farming system and other related factors. Therefore, the following are some drawbacks or challenges of traditional soil conservation methods in the study area.The results revealed that the farmers' experiences in traditional conservation practice were generally ample for any valuable evaluations. Major crops grown in the area include: maize, sugar-cane, rice, sweet potatoes, guinea corn and vegetables. Crop establishment in the locations followed compatibility to land use, soil texture and topography as much as the traditional conservation practices place (Ekwue&Tashiwa, 1992; Tekwa&Usman, 2006).

2.5.1 Socio cultural challenges

Old aged people who live in the study area are not involving in conservation practice due to their low potential to work. According to the survey made for this study, majority of household heads are aged old. Due to this, conservation practices like, soil or stone bunds in the kebele are adopted by small number of households. Besides, the sex of households in the kebelewill strong effect on adoption of heavy labor demanding conservation practices like soil and stone bunds. Therefore, this and other social related challenges are hampering the adoption of some of the physical conservation practices. These socio-cultural challenges have been modified and used in further studies – for instance, AlfonsTrompenaars conducted a comprehensive study, using the five factors developed by Parsons, adding time and environment as additional cultural dimensions (Trompenaars, Hampden-Turner 2005),

Due to the presence of traditional culture the farmers who lives in reduces the application of crop rotation for the need of short come income. Therefore, market price has strong impact in adoption of crop rotation in the kebele. Therefore, cultural challenges are heavily affecting the adoption of various physical or other conservation practices in the Keble.

3. RESEARCH METHODOLOGY

3.1 Description of the Study Area

Debre Work is a small town in western Ethiopia. Located in the East Gojjam of the Amhara Region, it has a latitude and longitude of 6°52′N 35°31′E with an elevation of 2489 meters above sea level. The settlement is known for its church and a hilltop monastery dedicated to the Virgin Mary, around which the community grew. It is the larger of two towns inEnarjEnawgaworeda.

An early mention of Debre Work is around 1620, when the Tulema Oromo were said to have devastated Gojjam as far as DebreWork.Its next mention is in 1684, in an itinerary of Emperor Iyasu I. Cardinal GuglielmoMassaia met with RasAli II here in the mid-19th century, describing the town as sitting on a round hill with the church at its summit; the town encircling the church almost extended down to the base of the hill. A debtera, who was head of the church, also appeared to be the civil head of the town. Charles Beke, who visited Debre Work 25 April 1842, described the town was not only located "on a conical eminence", located the hill at "the fork of two small streams Tazza and Zinjut." (Both of these streams are tributaries of the Chee.)

Paul B. Henze describes visiting the church at Debre Work in the early 1970s, an event which included finding the church being rebuilt from the foundations, and an unpleasant encounter with a monk over wanting to view the church's collection of manuscripts.

Demographics

Based on figures from the Central Statistical Agency in 2005, Debre Work has an estimated total population of 13,908, of whom 6,643 are men and 7,265 are women. The 1994 census reported this town had a total population of 8,048 of whom 3,495 were men and 4,553 were women.

3.2 Research design

The research design employed for the study was cross sectional survey type of research involving both qualitative and quantitative approaches were used. The reason for why Cross sectional research design is used is that, it involves using different groups of people who differ in the variable of interest but share other characteristics, such as socioeconomic status, educational background, and ethnicity. Qualitative data was from farmers understanding on line sowing technology through key informant survey, observation and other methods. Quantitative data was designed from sample farmers by using statistical methods and tools. Furthermore, the study was used explanatory survey research to opportunity and challenges of traditional soil conservation practice.

3.3 Sources and types of Data

The sources of data to use in this study are both primary and secondary data sources. The primary data was collect from agricultural extension expert, developmental agents and through interview and secondary data was gather from different relevant document related to learners, facilitators and education bureau. Quantitative data was collected using semi-structured questionnaires that were filled out through face-to-face and direct interview of respondents. Qualitative data were gathered from key informant and focus group discussion.

3.4 Data Collection methods

Quantitative data was collected using structured questionnaires that were filled out through face-to-face and direct interview of respondents. Qualitative data was gathered from key informant and focus group discussion. Primary data source was be used in order to get first-hand information from respondents and it helps providing information for specific purpose of addressing the problem at hand. So that questionnaires were prepared and were asked to sample respondents and to gathered necessary information for the study. Secondary data were gathered through reviewing documents, and reports from woreda agricultural and development bureau and also from published journal, website, newspaper and magazines.

3.5 Sampling Techniques and Sample Size

Multi-stage sampling procedure was followed to select sample respondents in the study area. At the first stage Debre worke woreda was selected purposively due to the potential area of soil conservation from the East Gojam zone. At the second stage two kebele were selected from fifteen kebeles of the woreda by using simple random sampling procedures. At the third stage the households of the both kebeles were stratified in to participant and non-participants of soil conservation. Finally, the households were selected using simple random sampling techniques. For the purpose of this study simple random sampling technique were employed to select the respondent. This is by considering of the distance, the second reason is since there is shortage of time and budget so it is difficult to conduct research in other far places. In the simple random sampling were used to select the sample household for household survey. These techniques were employed with the assumption of it avoids bias of representative and it create an opportunity to all people in the population equal chance of being selected. So from I start to this assumption I want to select one nearest areas. Such as, (Diyateba kebele and Dijumeti kebele) purposely.

Mixed multi stage sampling procedures were used to select sample respondents from Debirre work agricultural information. The two kebeles in Debirre work town is selects purposively; by considering its proximity, time and budget and transportation. And from 15Kebeles of Debirre work wereda, two Kebele was selects by using random sampling technique. For the purpose of this study simple random sampling technique was employed to select the respondent n this study the two centers Diyateba kebele and Dijumeti kebeles the total numbers of population expert facilitators. From each kebele I selected 44 farmers.

Then, the sample size is 88. Hence a total population of 1500 HH and 88 respondents were selected.

Slovene Formula: $n = \frac{N}{1+Ne^2}$

Where: n = sample size

N = total number of household in two kebele

e = margin of error (9%)

N=1500

N=1500

(1+1500(0.09)2=114

From the formula sample respondents from the two kebele would be 114 but due to time experience and budget constraints the sample respondents reduced to 88.

3.6 Data analysis

The collected data about challenges and opportunities of traditional soil conservation measure from primary and secondary sources are analyzed by using descriptive statistics and then presented by using tables, and percentages.

3.7 Hypothesis of Variables

Hypothesis is a situation in which independent variable influence dependent variable.

I described those variables on my own based on the discussion on this study.

3.7.1 Independent variables

Demographic factor likes;

Age on the other hand, it leads to experience in utilization of information and positively influences farmers "decision to adopt new technology to increase the productivity.

Age is one of the demographic factors that is useful to describe households and provide indication about the age structure of the sample population. Age is also used as an alternative for measuring farming experiences.

Education level of the respondent: The variable refers to level of formal education the respondent has attended. The educational level of the respondent increase, the greater is the possibility for farmers to become aware of the importance of new technology for utilization of agricultural activities. Thus, education levels have positive influence in the production of improved crops.

Marital status of respondent influence utilization to information married farmers sought information more than un-married due to desire to produce more for family consumption. The desire to produce more could lead to agricultural information seeking and use.

Sex of the household most of the land management practices require more labor force. Women are often faced with more labor constraints than male farmers and male-headed households. Hence, male headed households are expected to adopt the practices than female-headed households and undertake different land management practices, as better endowed with labor

Family size; The existence of low household family size negatively influences the activity of practicing agricultural information and household that has large family size influence agricultural information positively

3.6.2 Dependent variable

Gender can be defined as a set of characteristics, role and behavior pattern that distinguish women from men by socially and culturally and relation of power between them.

4. RESULTS AND DISCUSSION

4.1 Descriptive Statistics

This chapter comprises the study findings to be discussed under different sections, based on the objectives of the study. The level of challenges and opportunities of traditional soil conservation practices and its challenges with the measure undertaken to decrease the problem are analyzed in this chapter. At the last areas where soil conservation practice and reforestation activities have been undertaken, was compared with areas where soil-water conservation practices and reforestation have not been addressed. The overall results show that the challenges of soil and water conservation practices and measure undertaken to decrease the problems are discussed here under.

4.2 Socio-demographic characteristics

4.2.1 Sex, Age and marital status of Respondents

Table1: Sex, Age and Marital Status of Respondents

variable		Frequency	Percentage
sex	Male	**47**	53.4
	Female	41	46.6
	total	88	100
Age	15-25	**9**	10.2
	26-36	19	21.6
	>36	60	68.2
	total	88	100
Marital	Married	78	88.6
	Single	7	**7.95**
	Divorced	3	3.40
	total	88	100

Source: my Survey, 2020

As mentioned in the Table 4.1. Out of sampled respondents 53.4 % were male and the remaining 46.6 % were female. And 10.2% of respondent were in the aged between 15-25, about 21.6% of respondent were 26-36 and 68.2 % of respondents were aged above 36 years. As a result more than half percent of the households are above 36 aged .This indicates old aged persons were not that much in the traditional soil conservation practices. In addition to these 88.6 % of the respondent were married, 7.95 of respondents were single and the rest 3.40% were divorced(separated).This indicates that more than half of the study area of respondents were married.

Based on marital status, married have greater role towards conservation with compared to single. This is because of for keeping food service to their child and they involved or participated in any conservation measure either individually or groups. Since, single was depends on their family and their attitudes towards the soil conservation practice were less regarding to widow.

Table 2 Distance of farm from residence

Distance	Frequency	Percentage
Near by	38	43.18
0 -200m	23	26.13
201-300m	13	14.77
301-400m	9	10.22
>400	5	5.68
Total	88	100

Source: my Survey, 2020

A according to table (4.2)most of respondents said the distance of the farm land from residence is (43.18%)this indicates the farmers land are near to their homestead and so it leads to apply traditional soil conservation practices(like animal dung) efficiently and effectively .

Table 3: Educational status of respondents

Educational level	Frequency	Percentage
Illiterate	66	75
Read and write	15	17.04
5-8 grade	4	4.54
9-12 grade	2	2.27
Diploma graduate	1	1.13
Total	88	100

Source: my Survey, 2020

According to table (4. 2) 75 % of the respondents were illiterate ,17.04 % of respondents were can able to read and write, about 4.54 % of respondents were 5-8 grades, 2.27 % were 9-12 grade and 1.13% of respondents were diploma . This indicates that more than half of the study area respondents were illiterate.

Table 4: Family size of respondent

Family size	Frequency	Percentage
<3	15	17.04
3-8	68	77.27
Above 8	5	5.7
Total	88	100

Source: my Survey, 2020

As express in the above table (4. 3) 17.04 % of respondents have the house hold of below 3,77.27 % of respondents have a house hold size of 3-8 and 5.7 % of respondents have house hold size above 8 .This indicates above half of the respondents have the house size 3-8. Based on house hold size, many households have a greater role towards conservation with compared to households have smaller in size because of there is more division of labor.

Table 5: Income source of respondents

Income source	Frequency/number	Percentage
On farm	62	70.45
Off farm	17	19.31
Both	9	10.22
Total	**88**	**100**

Source: my Survey, 2020

As mentioned in the above table 70.45% of respondents can get their income from on farm (agriculture and related to agriculture), 10.22 % of respondents were get income from off farm (outside agriculture) And 10.22 % of respondents can get their incomes from both off farm and on farm. This indicates most of respondents can get their income from agriculture.

4.3. Nature, Type and problem of erosion in the study area

Now a day, soil erosion is the most common environmental problems in developing countries where agriculture is the dominant livelihood .It is one of the serious problem and environmental challenges affecting developing countries including Ethiopia Hudson(1992).This problem is more observed in areas where agriculture is predominantly based on rain fall. The study area is one among the area which have soil erosion has been challenging for the livelihood of the local people due to the fact that they depend on agriculture for their livelihood. According to the researcher's observation and respondent's response, erosion in the study area is becoming serious and including almost all types of accelerated erosion. By seeing these problems, the researcher initiated to assess opportunities and challenges of conservation measure.

Table 6: Types of common soil erosion in the study area

Types of erosion	Frequency	Percentage
Splash	2	2.27
Sheet erosion	13	14.77
Rill erosion	32	36.36
Gully erosion	41	46.6
Total	88	100

Source: my Survey, 2020

According to the above table (4. 5) the type of erosion which are common in the study area are rain splash erosion, sheet erosion, rill erosion ,and gully erosion .these all mentioned erosion types are related to the water erosion.

As observed in the study area ,the extent and severity is different based on magnitude ,intensity ,and impacts from all mentioned erosion ,the most sever type and predominantly occurring erosion types are rill and sheet erosion and in some cases and in some case gully erosion .Rill erosion occurs when the flow start ,concentrated ,and crate a defined flow path. It is most dangerous types of erosion in the study area that damage crops especially at the time of germination. These kinds of erosion are highly eroding the fertility of soil from the upper course of the river .This removal of soil from higher ground is affecting crop cultivation practices.

Now a day the extent and effect of erosion is decreasing time to time due to the community participation and having a great awareness towards those traditional conservation measures. The extent of erosion in

The study area is higher as compared to the surrounding Keble due to its topographic features .The extent of erosion in the study area is reaches higher during the rainy season especially in June and august. During this time sever erosion occurs in the study area especially on farm land.

According to the trend assessment of the study area, there was high rate of soil erosion, but this time the rate is decreased because of the involvement of local people conservation strategies or measures due to the increment of awareness of people about soil erosion, its effectiveness and improved conservation mechanisms.

4.4 Farmers perception on Cause of soil erosion in the study area

Table 7: Cause of soil erosion in the study area

Cause of erosion	Frequency	Percentage
Less fallowing period	9	10.22
High rain fall	38	43.18
Extensive cultivation	18	20.45
Deforestation	23	26.13
Total	88	100

Source: my Survey, 2020

According to the information from table ,the cause of soil erosion in the study area can be less fallowed period 10.22%,high rain fall 43.45 %, extensive cultivation 20.45 % and deforestation 26.13 % .From this ,high rain fall is a serious cause of erosion in the study area this is because the intensive rain fall sever during the month of June and July .Improper use of farm land, deforestation, over utilization of land resource are the major factors which cause soil erosion in the study area, where as high rain fall is a natural factor.

4.5 Effects of soil erosion in the study area

According to this finding the serious influence of both human and natural factors in the study area makes community farm land faced to the problem of erosion. This soil erosion hazards were emerge many effects in the study area. High population pressure, due to high population growth led to shortage of production. This finally results in decline of fertility and productivity. Generally accelerated soil erosion had numerous adverse effects on ecosystem of the area and Harms the life condition of human population, their domestic animals as well as their fauna and flora.

4.6 Traditional soil conservation methods being practiced in the study area

The Diyateba kebele people are known by traditional soil conservation practice. As noted by the kebele elder of Diyateba kebele began such practice in the earliest time in older to combat soil erosion since people have been practicing to protect the soil from loss.

There are also known as by its biological and structural soil conservation methods ,the biological methods are used to maintain soil fertility, such methods are crop rotation and application of animal dung .While the structural methods the one that mainly used for the purpose of controlling soil removal ,such as soil or stone bund, contouring.

Table 8: The type and extent of traditional soil conservation practices applied in the study area.

Types of conservation measure	Frequency			
	Adopters	**Percentage**	NonAdopters	**Percentage**
Crop rotation	31	35	57	65
Soil or stone buns	16	18	72	82
Animal dung	41	46	47	54

Source: my Survey, 2020

4.6.1: Soil bund

Soil bund is an embankment constructed from soil along the contour with water collection channels or basin at its upper side. These methods are used not only for control of erosion but also for growth of grass because of the farmer's plant at the bund. According to the above Table 4.7 soil bund in the study area applied only which is 18.2%, and 81.8 % were nonuse this traditional soil conservation practices this is due to the farmers complain of farm size challenges that means the farmers said it becomes narrowing of the farm land, and economic challenges (those farmers who lives in Diyateba kebele and Dijumeti kebele were not give consideration for traditional soil conservation because they require off farm incomes rather than on farm incomes).Development team in the watershed mainly advises the farmer to constructs this conservation structure but the farmers do not apply this method. Soil bund is effective in controlling soil loss, retaining moisture, and ultimately enhancing productivity of land (WFP (2005).

4.6.2: Animal dung

Animal dung is widely accepted organic fertilizer in the society of Diyateba kebele and Dijumeti kebele. Therefore, the community of the study area often used the natural fertilizer to increase the fertility of the soil thus it can possibly be said that natural fertilizer is widely accepted and widely used for stabilization and increment of soil fertility and moisture of soil among the farmers in the kebele. According to the field survey made for this study, about 46% of farmers living in this kebele adopt application of manure. Besides presence of greater proportion of female to male population in the kebele has contributed its own influence for the greater adoption of animal manure since female have greater relation to cows and their dung. Animal dung is mostly applied in the study area most likely due to there is sufficient amount of livestock size including cattle, sheep, goats, horse and hens, the community of the study area often used the natural fertilizer to increase the fertility of the soil this natural fertilizer is widely accepted and widely used for stabilization and increment of soil fertility and moisture of soil. Also the presence of greater average household size, in this kebele is also expected to contribute some for adoption of greater adoption of animal manure in the kebele since presence higher household size would yield greater labor size. This could possibly help for transportation of manure and other heavy labor intensive activities.

4.6.3 Crop Rotation

According to the Table 4.7, 35.2 % of respondents are user of crop rotation and 64.7% of respondents are non-users of crop rotation due to lack of awareness of the community and problems related to culture for example wedding. Crop rotation has many agronomic, economic and environmental benefits compared to mono culture cropping. Appropriate crop rotation increase organic matter in the soil, improve soil structure, reduce soil degradation, and can result in higher yields and greater farm profitability in the long term. It is used to control weeds and diseases.

4.7 Practices of soil and water conservation

To decrease the problem of SWC the development agents, and experts of the natural resource of the area advice farmers as it is not important to destroy the terrace bench. They also advise farmers, as it would be better to redevelop the existing terraces and supporting them with other soil fertility and SWC measures such as use of compost, farmyard manure, improved fallows, and use of grass strips in areas that are intensively cultivated, with high population density and high levels of land fragmentation.

The practiced structural soil and water conservations measure and soil fertility increment practice.

4.7.1 Traditional Cut-off Drains

A traditional cut-off drain is the most important and widely practiced structural indigenous soil conservation measure in the study area. In the focus group discussions and key informant interview, farmers were asked on how to and when to construct the structure. They said that the structures were constructed across the traditional waterways, which were prepared on farm plots parallel to the slope during rainy season. Farmers have the intension of avoiding seed loss mainly cereal crops like „teff" and to increase yield per unit area (production-oriented) and to conserve soil from the erosive power of runoff (protection-oriented). Farmers of more villages of sample study area whose farmlands threatened by runoff do practice this traditional soil-water conservation method.

4.7.2 Traditional waterways

This structure is a widely practiced form currently throughout the country to
reduce soil loss by runoff on higher altitude

It looks like cutoff drains but it relatively covers wider area at the top of a hillside to drain runoff coming from any direction to the nearest riverbank. The structure are constructed mainly by neighborhood peasant house hold who own neighbor farm plots with the family labor as an input or by the farmers in group in the form of campaign. In the group discussions, farmers also emphasized the advantage of working in groups are that the constructed structure can serve for a long time. The stone paved waterways were used for a long period than the grass waterway, which was eroded by runoff. By doing so, they protect their farm plots from the damaging effect of runoff.

4.7.3 Drip irrigation

Drip irrigation, also known as trickle irrigation or micro irrigation or localized irrigation, is an irrigation method that saves water and fertilizer by allowing water to drip slowly to the roots of plants, either onto the soil surface or directly onto the root zone, through a network of valves, pipes, tubing(hose) , and emitters. It was done through narrow tubes that deliver water directly to the base of the plant form the „*tanker*", which the DA"s and farmer use, as source of water in the study area.

4.8 Challenges of Traditional Soil Conservation Practices in the Study Area

Traditional soil conservation practices have several benefits for the conservation of environment as a general and soil resource in particular .Even though such positive aspects and crucial value of conservation such as erosion control, maintenance of soil fertility and water conservation, but there are some limitations or challenges that have been observed in the study area This challenges may results from improper traditional farming system and other related factors. Therefore, the following are some drawbacks or challenges of traditional soil conservation methods in the study area.

4.8.1 Socio cultural challenges

Social Challenges

Old aged people who live in the study area are not involving in conservation practice due to their low potential to work. According to the survey made for this study, majority of household heads are aged old. Due to this, conservation practices like, soil or stone bunds in the kebele are only adopted by 18% of the total households. Besides, the sex of households in the kebele has strong effect on adoption of heavy labor demanding conservation practices like soil and stone bunds. Practically investigating about 87.23% of households in the kebele are male leaded households and on the other hand the soil or stone bund practiced in the kebele is only 18%. Therefore, this and other social related challenges are hampering the adoption of some of the physical conservation practices.

Cultural challenges

Due to the presence of traditional culture the farmers who lives in Diyateba kebele and Dijumeti kebele reduces the application of crop rotation for the need of short come income. Therefore, market price has strong impact in adoption of crop rotation in the kebele. One of the respondents in the kebele said that

"We continually practice or cultivate crops which are having high market demand rather than rotating crops having lower demand in the market. For instances, if there is weeding we already know in the kebele or nearby kebele, we prefer to cultivate Teff other than other crops due to its higher demand in market".

Also, according to the survey made so far for this study, 100% of respondents are orthodox Christianity fellows which on the other hand imply as there is much more missing days for agricultural practices.

Therefore, cultural challenges are heavily affecting the adoption of various physical or other conservation practices in the Keble.

4.8.2 Deforestation

As survey of this study indicates the outcome was a mass movement of topsoil down the slope during the main rainy season leading to flooding and sedimentation at the valley bottom. The hills also failed to retain soil and water leading to the drying up of the springs at the foothills. Overgrazing has further contributed to the land degradation exposing the soil to wind erosion during the dry season. Hence, erosion and land degradation can bring a total change of the agro-biodiversity to a village, e.g. diversity in soil organisms, crop and livestock genetic diversity, and wild life, at medium and lower elevations from time to time and caused severe water and wind erosions mostly in northern and eastern part of the district.

4.9 Opportunities of Traditional soil conservation practices in the study area

Traditional soil conservation structures are not only to control run off and soil erosion as in the case of newly introducing measures, but also for the production of high quality and quantity yield for community. In Diyateba and Dijumeti kebele there are many traditional soil conservation practices those are both biological practice like application of animal dung, crop rotation and mechanical measures like soil or stone bund .The farmers that live in Diyateba kebele uses these conservation practices because of there are opportunities that are suited for them . Suitable conditions that the farmers applying these measures are;

Presence of livestock

As the information obtained from the respondents those farmers have large number of livestock size have more attention to traditional soil conservation practices. Because there is enough amount of dung due to presence of large livestock in each family, the opportunity of farm land for being managed properly is likely. Therefore, amount of livestock has great importance for adoption of manure as traditional soil and water conservation practices.

Labor

The application need extensive labor to take dung from home to farm land and it needs excessive time for collecting martial at the time of preparation. So it affects the local community who lives in study area.

Distances of farm from residence

Distance of the farm from the homestead is one opportunity for farmers to apply those traditional soil conservation practices. As obtained the information from respondents in the above Table 2 most of the respondents (43.18%) said where the farm land is nearest to the homestead have the chance to conserve properly than that of that is located far. So in general the farm land which is far from the homestead less likely to apply those traditional soil conservation practices due to the consideration of labor.

Cost of inorganic fertilizer

The cost of inorganic fertilizer becomes increase time to time. As the researcher obtained the information from respondents those farmers that are present in the study area are apply traditional soil conservation practices like application of animal dung, crop rotation (of the leguminous trees that increases soil fertility) in order to reduce expense of inorganic fertilizer. According to the interview with one of the farmer in the Keble, the price of UREA as well as DAP is increasing from time to time. And the increment of price of inorganic fertilizer in return has declined farmer's preference of using inorganic fertilizer.

5 CONCULISION AND RECOMMENDATION

5.1 CONCULISION

This study was conducted to assess the opportunities and challenges of traditional soil conservation methods on soil erosion in Amhara regional state, in the case of Diyateba keble and Dijumeti Keble.

According to the analysis made so far in this research, females have equal responsibility in the Keble including conservation of land. Besides, age of the household heads is under category of old age which in return has greater influence on adoption of heavy labor demanding conservation practices. Literacy issue in the Keble is so worst that majority of residents of Keble are illiterate and the researcher found as it has great impact on conservation of soil and water conservation technologies.

The people of the study area have been practicing traditional soil conservation methods for decades and even more than that. These soil conservation methods are both biological and structural soil conservation measures. However, these traditional soil conservation methods have got their own challenges and opportunities to be adopted by the farmers.

Among the challenge need for extensive labor, economic challenges, and cultural challenges on traditional soil conservation methods are the major one.

While regarding opportunities of traditional soil conservation methods includes, labor, presence of available resource, cost of inorganic fertilizer and distance of the farm land from the residues.

The people of the study area found to both methods, but the biological one is found to be dominant.

From the information that we obtained most of the respondents in the study area are not very much to continue with traditional soil conservation methods.

Generally, as we observe there are a number of challenges which hinder traditional soil conservation practices to control soil erosion as well as opportunities.

5.2 Recommendation

That any issues need solutions to overcome its drawbacks and some is true, in this study some gaps require suggestion or recommendation. Therefore, based on the finding made, the following recommendations are made.

- The most common type of erosion in the study area is gully erosion, so there should be control method through check dams, wood brush or other scientifically proved soil and water conservation technologies.
- Government has to work hardly towards in reducing or avoiding bad cultures or extravagant activities through awareness creation or any mass campaign.
- Government has to consider both off farm and on farm households through better time allocation mechanism so that all involvers can act accordingly.
- The focuses on this study was mainly on socio economic challenges, therefore, the researcher recommends further researchers to deal on technological and farm institutional factors.

6. REFERANCES

B adage B (2001) Deforestation and land degradation in Ethiopian high lands astrategy for physical recovery .Norhern african studies ,ISSN 0740-9133,8(1):7-26.

Babul B ,muys B ,Nega F. Tollens E ,Nyssen J ,Deckers J, Malthijise (2009) The economic contribution of forest resources used for rural livelihood in Ti gray ,North en Ethiopia .Forest policy and economics ,11(2):109-117.

Balana ,B .B ,Mathijis ,E. s and Muys ,B(2010) Assessing the sustainability of forest management :an application of multi criteria decision analysis to community forest in Northen Ethiopia J .Envirn manage :91,1294-1304.

Bishaw ,B (2001) Deforestation and land degradation in Ethiopian highlands : A stratagy for physical recovery . North east african studies ,8(1) 325-358.

Ervin ,C.A .and D.E Ervin ,(1982) Factors affecting the use of soil conservation practices :Hypothesis ,evidence and policy implementation .Land Econ.,58(3):277-292.

FAO / World bank,(2001) Farming system and poverty : Improving livelihoods in changing world. Washington. D.C ,:Food and Agriculture organization of the united nations (FAO) and world bank.

Grma ,T ,(2001) Land degradation : a change to Ethiopian journal of environmental management ,27(6) :815-824.

Hudsen,N.,1992 .Soil conservation . Bats fored ;Limited campanies.Inc.New York.

Mekuria W,.,veldkample ,tilahun M,olshewsk:R (2011) Economic valuation of land restoration. Land degradation aqnd development ,22(3):334-344.

Morgan ,R.P.C., (2005) Soil and water conservation .2nd edition .Blackwell pub.carnfield university.

Nyssen J,Poeson J,Deckers J (2009) Land degradation and soil and water conservation in tropical highlands .Soil and Tillage Research ,103(2):197-202.

Rafahi ,H,(2004) Control of water erosion .Tehran university press.

Reij,C.,I .Scoones and C .Toulmin ,(1996).Sustaining the soil :indigenous SWC in Africa .Earthscan,publication ltd ,Londen

Rogress,E.M.,(2003) . Diffusion of innovation ,5th Edn. New york :Free press.

Senait (2005) Determinant of choice of land management practice :A case study of Ankober District ,Ethiopian journal of Agriculture service ,18(1):53-67.

Stark ,M.,(2000). Soil management strategy to sustain continuous crop production between vegetative contour strips on humid tropical hillsides. International Center for Research in Agroforestery ,Southeast Asia.

Tavakoli ,M,Y,Mohammadi and E,piri ,2006 . The effect of implementing pasture management projects in soil conservation in Iran province presented 3rd national conference of erosion and sediment.Teharan.

Wagayehu,B (2003) . Economics of Soil and Water Conservation . Theory and empirical application to subsistence farming in the Eastern Ethiopia highlands ;PhD Thesis ;Swidish University of Agricultural Science . Uppsala 2003.

Yitayal ,A.(2004) Determinant of use of conservation measures by small holder farmers :In Gimma zone ,Debo District.M.sc .Thesis presented to the school of graduate studies of Alamaya university .pp.65-82.

Yitbarek ,T.W,Bellietheran ,S .,and stringer ,L.C .(2012) The on site cost of gully erosion and cost-benefit of gully rehablitation : a case study in ethiopia .Land Degra.Dev.23,157-166.

Yohannes G.M and Herweg,K .(2000) Farm indiginouse knowledge to participatory technology development :Centere for Development and Enivoronment (CDE) ,University of Bern.